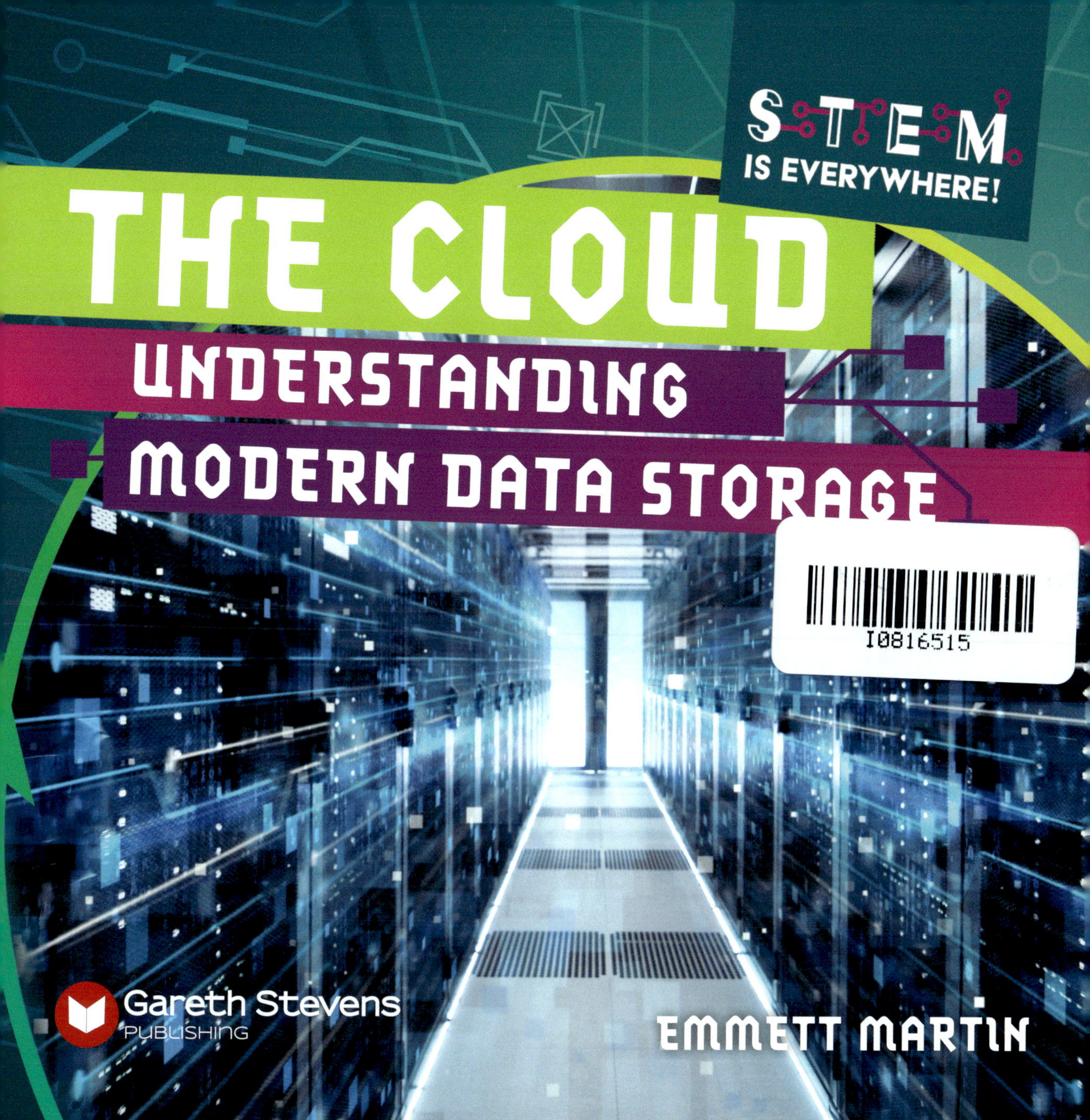
STEM
IS EVERYWHERE!
THE CLOUD
UNDERSTANDING
MODERN DATA STORAGE
I0816515
Gareth Stevens
PUBLISHING
EMMETT MARTIN

Please visit our website, www.garethstevens.com.
For a free color catalog of all our high-quality books, call toll free 1-800-542-2595 or fax 1-877-542-2596.

Portions of this work were originally authored by Jeanne Marie Ford and published as *How the Cloud Works* (Everyday STEM). All new material in this edition authored by Emmett Martin.

Library of Congress Cataloging-in-Publication Data
Names: Martin, Emmett, author.
Title: The cloud : understanding modern data storage / Emmett Martin.
Description: New York : Gareth Stevens Publishing, [2023] | Series: Stem is everywhere! | Includes bibliographical references and index.
Identifiers: LCCN 2022026060 (print) | LCCN 2022026061 (ebook) | ISBN 9781538283653 (library binding) | ISBN 9781538283639 (paperback) | ISBN 9781538283660 (ebook)
Subjects: LCSH: Cloud computing–Juvenile literature.
Classification: LCC QA76.585 .M384 2023 (print) | LCC QA76.585 (ebook) | DDC 004.67/82–dc23/eng/20220729
LC record available at https://lccn.loc.gov/2022026060
LC ebook record available at https://lccn.loc.gov/2022026061

Published in 2023 by
Gareth Stevens Publishing
2544 Clinton Street
Buffalo, NY 14224

Designer: Tanya Dellaccio
Editor: Therese Shea

Photo credits: Series Art Supphachai Salaeman/Shutterstock.com; Cover Gorodenkoff/Shutterstock.com; p. 5 (top) Pakhnyushchy/Shutterstock.com; p. 5 (bottom) Blackboard/Shutterstock.com; p. 6 Jiri Hera/Shutterstock.com; p. 7 https://upload.wikimedia.org/wikipedia/commons/c/ce/Xbox-360-512MB-MemCard.jpg; p. 9 (bottom) Overdose Studio/Shutterstock.com; p. 9 (top) https://upload.wikimedia.org/wikipedia/commons/8/8c/Altair_8800_-_MfK_Bern.jpg; p. 11 Pixel-Shot/Shutterstock.com; p. 12 https://upload.wikimedia.org/wikipedia/commons/6/66/J._C._R._Licklider.jpg; p. 13 https://upload.wikimedia.org/wikipedia/commons/b/bc/Arpanet_in_the_1970s.png; p. 15 FellowNeko/Shutterstock.com; p. 17 Oleksiy Mark/Shutterstock.com; p. 21 (Twitter and Facebook logos) fyv6561/Shutterstock.com; p. 21 TikTok and Instagram logo) rvlsoft/shutterstock.com; p. 21 (main) Dan Rentea/Shutterstock.com; p. 22 Ground Picture/Shutterstock.com; p. 23 Andrey_Popov/Shutterstock.com; p. 25 matejmo/iStock.com; p. 27 Itsanan Sampuntarat/iStock.com.

Printed in the United States of America

CPSIA compliance information: Batch #CWGS23: For further information contact Gareth Stevems Publishing at 1-800-398-2504.

CONTENTS

A CLOUDY IDEA 4
HOME OR AWAY? 6
BEFORE THE CLOUD 8
CREATING THE CLOUD 12
WIDE ACCESS 14
CLOUD TO CLOUD 16
THE CLOUD BRINGS CHANGE 18
APPS IN THE CLOUD 20
THE CLOUD IN THE PANDEMIC 22
CLOUD CONCERNS 24
AI AND THE CLOUD 26
THE EXPANDING CLOUD 28
GLOSSARY 30
FOR MORE INFORMATION 31
INDEX 32

Words in the glossary appear in **bold** type the first time they are used in the text.

A CLOUDY IDEA

In weather, a cloud is a collection of **billions** of tiny pieces of water or ice. Up in the air, mixed with bits of dust, it can look like a huge puffy white mass. In **technology**, the cloud is completely different. It isn't located in the sky. And it isn't water or ice—it's data.

The cloud is made up of programs, services, and other kinds of data reached through the internet. It isn't anything you have stored on your own computer, tablet, smartphone, or other smart device. Instead, it's a kind of online storage. In this book, you'll learn how the cloud has changed the way we live.

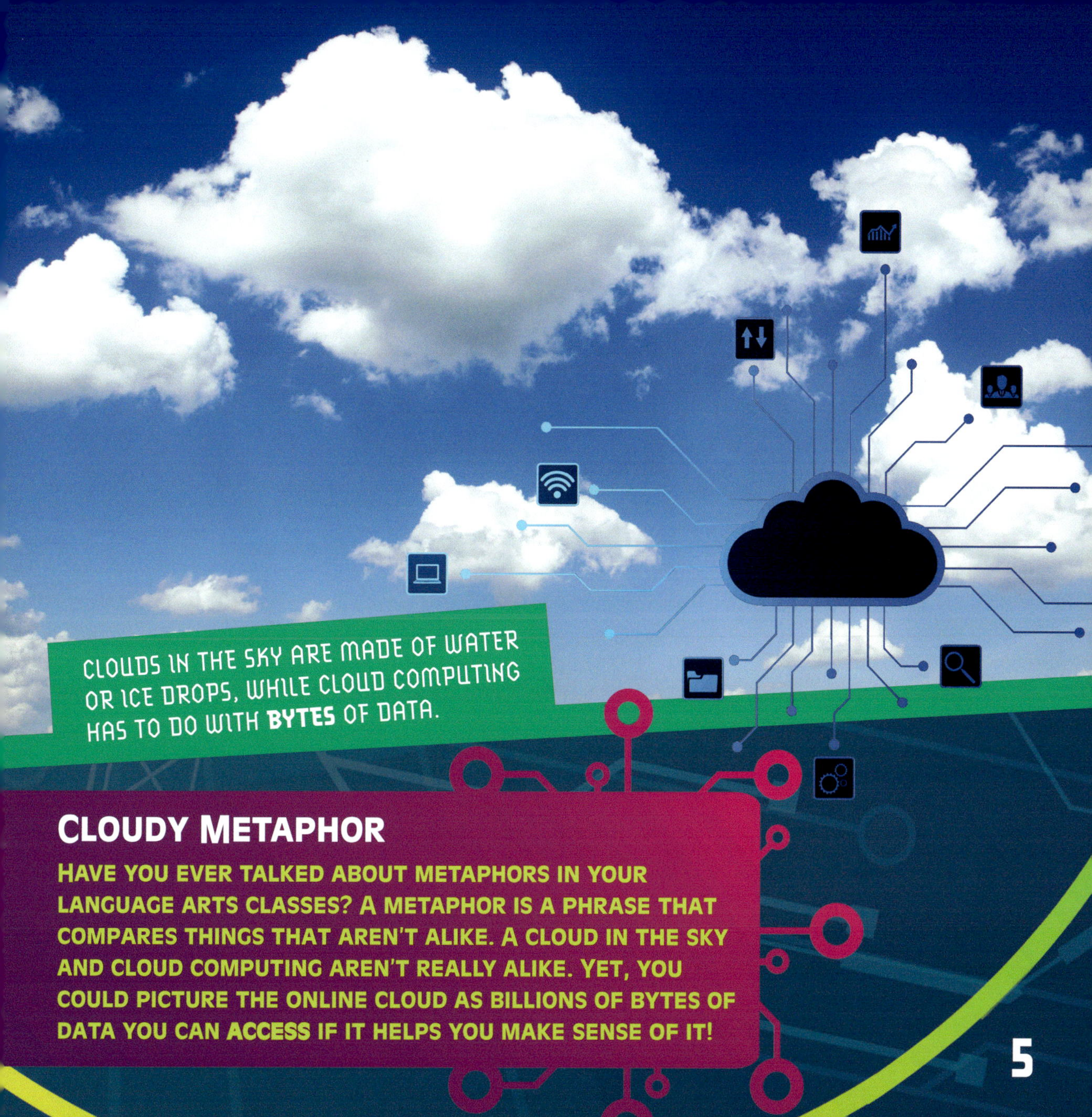

CLOUDY METAPHOR

HAVE YOU EVER TALKED ABOUT METAPHORS IN YOUR LANGUAGE ARTS CLASSES? A METAPHOR IS A PHRASE THAT COMPARES THINGS THAT AREN'T ALIKE. A CLOUD IN THE SKY AND CLOUD COMPUTING AREN'T REALLY ALIKE. YET, YOU COULD PICTURE THE ONLINE CLOUD AS BILLIONS OF BYTES OF DATA YOU CAN **ACCESS** IF IT HELPS YOU MAKE SENSE OF IT!

HOME OR AWAY?

Information, or data, saved on your own devices is called local storage. Hard drives and memory cards are local storage. They hold limited amounts of data. Cloud servers also save information. However, they have much larger memories. They allow users to store far more data online. Today's computers, smartphones, and tablets use cloud technology for storage.

Many **apps** are cloud-based. Messenger is one popular cloud-based app in which users can send messages, videos, and other files to friends and receive them too. This data isn't saved on users' devices. It's in the cloud—it can be accessed on any device with the Messenger app and an internet connection.

A Streaming Society

MANY PEOPLE ACCESS THE CLOUD MANY TIMES EVERY DAY. IF YOU'RE SOMEONE WHO STREAMS TV SHOWS, MOVIES, OR MUSIC, YOU'RE ACCESSING THE CLOUD. BY STREAMING, YOU'RE USING A TECHNOLOGY THAT SENDS A STREAM OF DATA FROM A SERVER ONLINE TO YOUR SMART DEVICE. IT DOESN'T STORE THE FILE IN YOUR DEVICE'S PERMANENT, OR LASTING, MEMORY.

THESE ARE SOME KINDS OF LOCAL STORAGE DEVICES, INCLUDING MEMORY CARDS FOR A VIDEO GAME SYSTEM.

BEFORE THE CLOUD

Servers that store information for cloud-based computing are powerful computers built to manage large amounts of data. These machines are much different than the first computers, which could fill a room. But companies back then paid to share computer time to save money. This idea of sharing computer **resources** led to the cloud.

Over time, computers got smaller and more powerful. The first personal computers (PCs) were sold in the 1970s. Their hard drives could store only a small amount of data at first. PC users could save more data on external drives, or devices that can be plugged into a computer port for more storage.

THE ALTAIR WAS THE FIRST WIDELY SOLD PERSONAL COMPUTER. IT'S SHOWN HERE WITHOUT A KEYBOARD.

THE FLOPPY DISK

EARLY PC USERS COULD SAVE THEIR DATA TO WHAT WAS CALLED A FLOPPY DISK. IT WAS MADE OF FLEXIBLE, OR BENDABLE, PLASTIC COATED WITH MAGNETIC MATTER. THE DISK WAS PLACED INTO A COMPUTER'S FLOPPY DISK DRIVE, WHERE AN **ELECTROMAGNET** RECORDED OR READ DATA ON THE DISK AS THE DISK SPUN.

Local storage like floppy disks and, later, **flash drives** had problems, though. Data could become corrupted, making the files unable to be accessed. Disks and flash drives could also be lost or left somewhere. For example, students needed to bring them to school or home, or wherever they wanted to access them. The cloud makes it easier to move, share, and access data from anywhere.

The cloud solved another problem too. Games and music could fill up all the storage on a PC. But cloud storage means you don't have to store large files on your computer. They can be stored and accessed in the cloud.

CLOUD GAMING IS PLAYING GAMES USING AN ONLINE STREAMING SERVICE RATHER THAN DOWNLOADING THE GAME ONTO A DEVICE.

WHAT IS CORRUPTION?

DATA CORRUPTION MEANS INFORMATION IN A FILE OR PROGRAM IS LOST OR HARMED ACCIDENTALLY AS IT'S BEING RECORDED OR ACCESSED. IT'S NO LONGER READABLE. THIS CAN HAPPEN IF THERE'S AN INTERRUPTION DURING A DATA PROCESS, LIKE THE POWER GOING OUT. CORRUPTION CAN HAPPEN ON ANY DEVICE, INCLUDING ON A DISK OR IN A HARD DRIVE. COMPUTER VIRUSES ARE ANOTHER WAY DATA IS CORRUPTED.

CREATING THE CLOUD

Joseph Carl Robnett Licklider (shown below) is one of the founders of cloud computing. In 1962, he began writing about his idea for a giant computer network. He pictured that his network would stretch across the world. He **predicted** many things that would happen with computers.

SEEING INTO THE FUTURE

COMPUTER SCIENTIST JOHN MCCARTHY PICTURED THE CLOUD IN 1961. HE IMAGINED IT AS A PUBLIC UTILITY, OR SERVICE OVERSEEN BY GOVERNMENT RULES, SUCH AS A TELEPHONE OR ELECTRIC COMPANY. HE THOUGHT USERS COULD PAY TO USE AS MUCH DATA AS THEY NEEDED. TODAY, THE CLOUD IS SET UP JUST LIKE THIS.

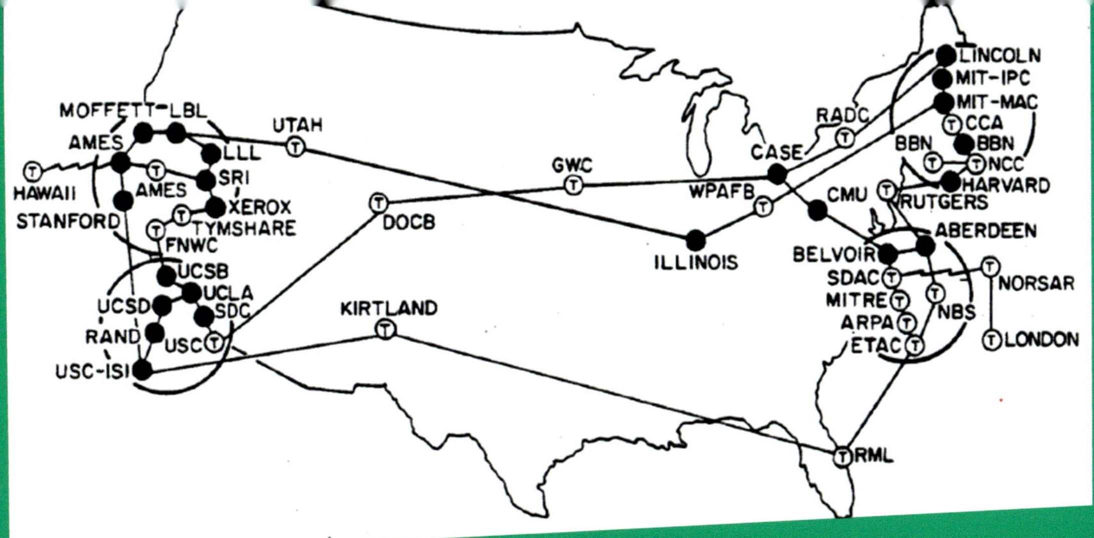

ARPANET GREW LARGER THROUGH THE 1970s, AS THIS MAP OF ACCESS POINTS SHOWS. THE U.S. GOVERNMENT GREW WORRIED TOO MANY PEOPLE WERE USING IT. THEY SPLIT IT INTO TWO: A MILITARY NETWORK (MILNET) AND ARPANET.

In the 1960s, Licklider's ideas inspired a computer network called ARPANET (Advanced Research Projects Agency Network). At first, it connected four university computers: the University of California, Los Angeles; the Stanford Research Institute in Menlo Park, California; the University of California, Santa Barbara; and the University of Utah. ARPANET was shut down in 1990, but the technology behind it made the internet possible.

WIDE ACCESS

The first personal computers didn't have internet access. Interest in the internet grew in the 1990s. At first, internet data moved over phone lines. It traveled slowly and people couldn't send large files. Today, we have high-speed internet that can be accessed in many ways. A lot of data can be transmitted very quickly.

High-speed internet made cloud storage possible. Cloud servers can be accessed from anywhere with an internet connection. Today, many personal computers use the cloud for storage. This allows them to be lighter and less costly. Tablets, smartphones, and other smart devices also use cloud apps.

How We Connect

Internet connections depend on an internet service provider (ISP). Some ISPs offer high-speed fiber-optic internet, which uses cables with flexible glass or plastic strands through which light signals can be sent. Some services connect to the internet through satellites. Cellular internet services are a wireless connection for our smartphones and other smart devices using networks of cell towers.

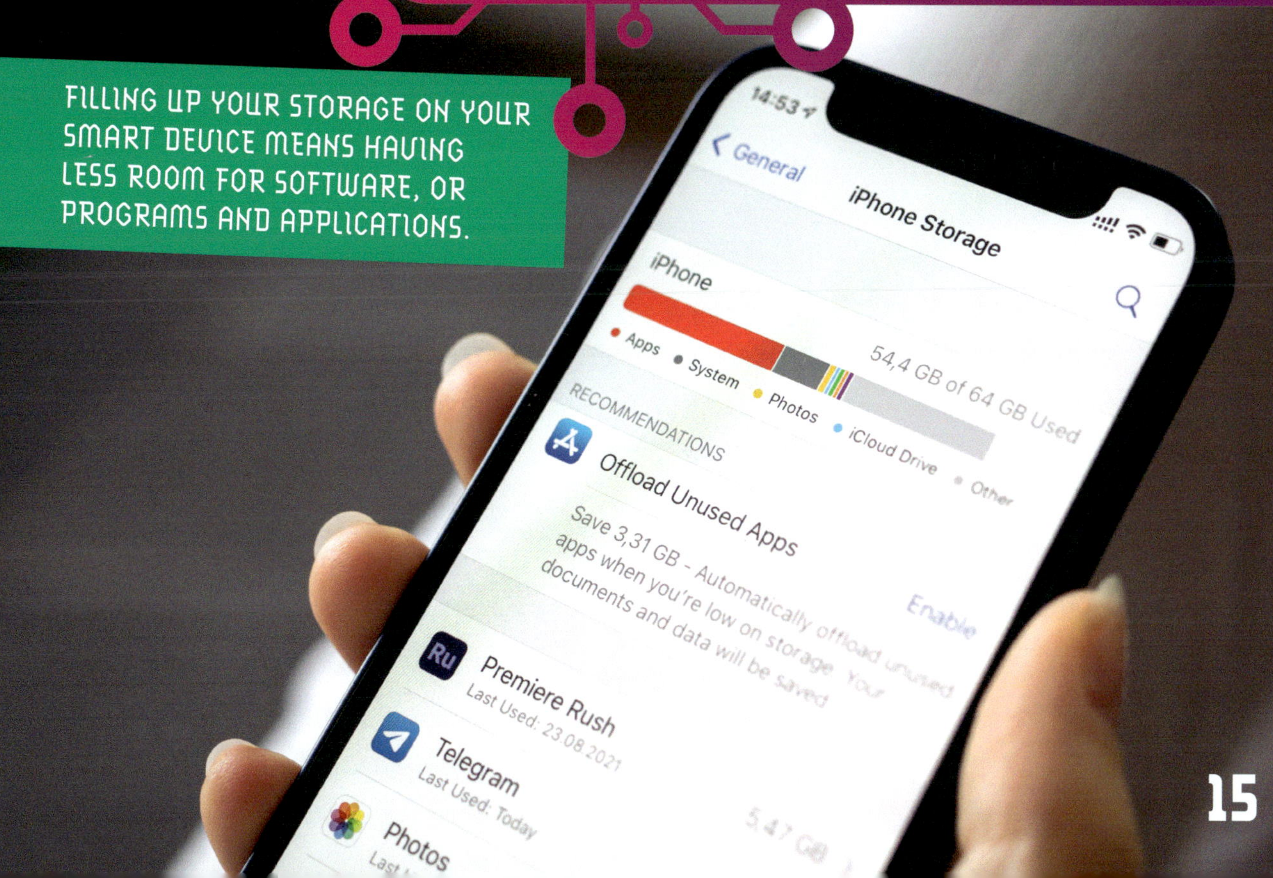

FILLING UP YOUR STORAGE ON YOUR SMART DEVICE MEANS HAVING LESS ROOM FOR SOFTWARE, OR PROGRAMS AND APPLICATIONS.

CLOUD TO CLOUD

The cloud isn't just one network but many. Some networks are available to the public so everyone can use them. Some are owned by companies and are private, used only by employees of that company. Some networks are called hybrids. Hybrid clouds carry both public and private networks, for example for customers and employees. All clouds work in the same basic way, though.

SERVER VS. COMPUTER

A SERVER IS A COMPUTER, BUT UNLIKE OUR PERSONAL COMPUTERS THAT DO MANY DIFFERENT TASKS, SERVERS ARE MAINLY MEANT TO STORE, MANAGE, SEND OUT, AND RETRIEVE DATA WHEN IT'S NEEDED. SERVERS NEED TO BE ABLE TO PROCESS REQUESTS QUICKLY AND HAVE A LOT OF STORAGE. MANY RUN 24 HOURS A DAY.

LARGE CLOUD-BASED COMPANIES NEED A LOT OF SERVERS. THEY MIGHT STORE THEM ON RACKS IN A SERVER ROOM LIKE THIS.

A local network connects computers within the same business or location. Switches let these computers communicate. **Routers** and **modems** link local networks and let them send and receive data over the internet. Servers all over the world store and share data with networks.

THE CLOUD BRINGS CHANGE

Cloud technology has changed work life. Employers can store and access data on the cloud, without having to buy their own servers or even software. Companies can share resources, buy less software, and save money on technology. They can then spend money on other things that help their businesses grow.

WORKING AT HOME

CLOUD USERS CAN WORK FROM HOME MORE EASILY. COMPANIES SAVE MONEY ON OFFICE SPACE WHEN EMPLOYEES WORK AT HOME. OFFICES DON'T HAVE TO BE AS LARGE. WORKING FROM HOME HELPS EMPLOYEES SPEND LESS TIME COMMUTING. IT CAN LESSEN THE NEED FOR COSTLY BUSINESS TRIPS.

SMALL BUSINESSES JUST STARTING OUT CAN USE CLOUD SERVICES FOR SOFTWARE AND HARDWARE, OR COMPUTER DEVICES LIKE SERVERS. THEY PAY, SOMETIMES MONTHLY, BUT LESS MONEY THAN IT WOULD COST TO BUY AND MANAGE SOFTWARE AND HARDWARE.

THE THREE CLOUD-COMPUTING SERVICES

IAAS (INFRASTRUCTURE AS A SERVICE)

ACCESS CLOUD-MANAGED RESOURCES SUCH AS SERVERS, NETWORKS, AND STORAGE, INSTEAD OF PAYING FOR AND MANAGING HARDWARE

EXAMPLE: SETTING UP AN ONLINE BUSINESS

PAAS (PLATFORM AS A SERVICE)

ACCESS CLOUD-BASED SOFTWARE AND TOOLS IN ORDER TO BUILD NEW APPS AND SOFTWARE

EXAMPLE: DEVELOPING A SMARTPHONE APP

SAAS (SOFTWARE AS A SERVICE)

ACCESS CLOUD-BASED SOFTWARE OVER THE INTERNET RATHER THAN DOWNLOADING THE SOFTWARE

EXAMPLE: USING WEB-BASED EMAIL

Sharing files through the cloud speeds up processes. Doctors can look at medical records from another office, even in another state, to help their patients much more quickly. People working together can exchange data almost instantly. For example, two or more employees can edit files at the same time. This is called synchronization. Synchronization makes information match.

APPS IN THE CLOUD

Many of the most popular applications you might know use cloud-based technology. Twitter, Instagram, TikTok, and Facebook are cloud-based apps. Cloud-based apps operate at least partly through the internet. You may need to download something to your device, for example an app to access your account, but data storage and processing are done through the cloud. No matter where you are, you can access your account with a smart device.

Cloud-based apps have changed the way we communicate with our friends. They are how some people keep up with loved ones and share news—or make new online friends around the world!

CLOUD-BASED APPS PROVIDE ENTERTAINMENT AND A WAY TO ENTERTAIN OTHERS!

AGE APPROPRIATE

MANY APPS REQUIRE A MINIMUM AGE TO SIGN UP FOR AN ACCOUNT. FOR FACEBOOK, THAT AGE IS AT LEAST 13 YEARS OLD (DEPENDING ON WHERE THE USER LIVES). THE TIKTOK SHORT-VIDEO APP OFFERS TIKTOK FOR YOUNGER USERS FOR THOSE UNDER 13. BUT ALWAYS CHECK WITH A TRUSTED ADULT BEFORE YOU DO ANYTHING ON THE INTERNET.

THE CLOUD IN THE PANDEMIC

The advantages of cloud-based technology became more noticeable during the COVID-19 **pandemic**. Many people stopped going into offices to work. Cloud-based technology allowed them to continue their jobs at home. Some businesses changed how they operated, allowing employees to continue working from home even after they could safely go back to the office.

CLOUD-COMPUTING HAS HELPED STUDENTS KEEP UP WITH THEIR CLASSWORK, EVEN IF THEY'RE LEARNING FROM HOME.

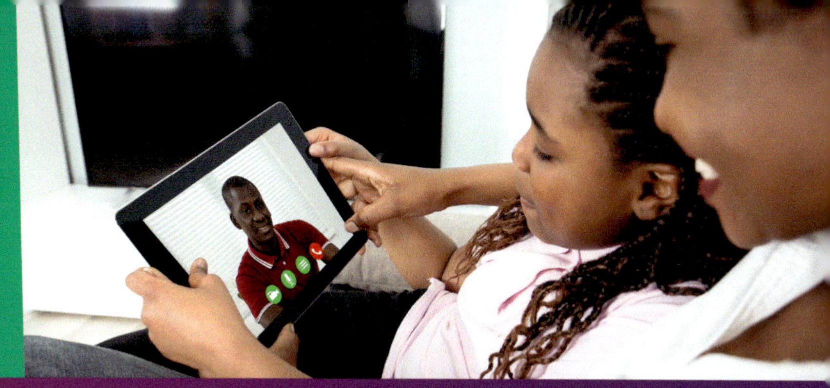

Meet Me in the Cloud

During the COVID-19 pandemic, cloud-based apps like Zoom and FaceTime helped people keep in touch with their families and friends while staying safe. Some people got creative and even played games or watched movies over the internet together—while miles apart. In December 2020, it was reported 350 million people used Zoom every day!

Cloud-based technology also allowed students to continue their education at home. Schools used apps such as Google Classroom and Zoom to help students communicate with their teachers and each other, receive and hand in assignments, and to track their grades. Like workplaces, many schools continued to use this technology after students returned to the classroom.

CLOUD CONCERNS

The cloud does have drawbacks, however. Cloud users cannot access data without the internet. So people who don't have a connection to the internet for any reason, such as a lack of money, their location, or a power outage, cannot access whatever service or information they're looking for.

Keeping private data secure is another issue of concern. Hackers, or people who access computer systems illegally, can steal private information, like credit card numbers. Encryption of data in the cloud makes it more secure. Encryption uses codes so hackers can't read the data. Computer scientists are always trying to make online storage safer.

DATA ENCRYPTION CHANGES INFORMATION INTO A KIND OF CODE. ONLY PEOPLE WITH A "KEY" OR A PASSWORD CAN ACCESS IT IN THE CORRECT FORM.

INTERNET FOR ALL

OFTEN PEOPLE CAN ACCESS FREE WIRELESS INTERNET IN PLACES SUCH AS PUBLIC LIBRARIES AND CAFÉS. MANY CITIES ARE MAKING EFFORTS TO PROVIDE INTERNET SERVICE FOR ALL PEOPLE WHO LIVE THERE TOO. IN 2022, IT WAS ANNOUNCED THE U.S. GOVERNMENT UNDER PRESIDENT JOE BIDEN PLANNED TO SPEND $45 BILLION TO PROVIDE INTERNET ACCESS TO AMERICANS ACROSS THE NATION.

AI AND THE CLOUD

Cloud computing is expected to keep growing. Part of this growth may be in artificial intelligence (AI). AI is the ability of a computer or digital device to solve problems and make decisions like a human being would.

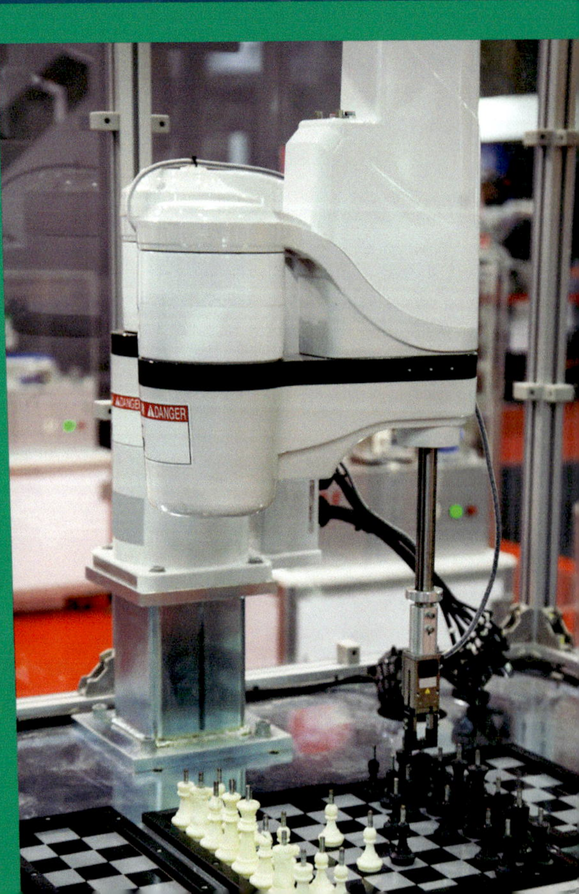

THIS ROBOTIC ARM USES AI TO PLAY CHESS.

LOOKING AHEAD

THE BIGGER THE CLOUD GETS, THE MORE DATA IS AVAILABLE FOR AI TO BECOME MORE ADVANCED. IN THE FUTURE, AI MAY BE ABLE TO WATCH OVER SYSTEMS AND PROCESSES WITHIN THE CLOUD AND HELP FIX PROBLEMS, JOBS THAT COMPUTER SCIENTISTS NOW DO. AND MORE AND MORE BUSINESSES WILL BE ABLE TO USE AI THROUGH CLOUD-BASED SERVICES.

You might already use AI, through a digital assistant. A digital assistant is a kind of computer program that can talk with someone. It uses AI and other processes to answer questions and do helpful tasks. Cloud technology gives AI programs access to answers and services for the user. Amazon's Alexa, Apple's Siri, and Google Assistant are digital voice assistants.

THE EXPANDING CLOUD

Cloud technology has grown so much in just a few years. It does have a cost, though. In 2022, people spent hundreds of billions of dollars to use cloud services, and this amount is expected to increase. In 2021, cloud-based video streaming alone was worth over $59 billion and is expected to grow even more profitable.

What do you do on the cloud: email, homework, look at photos, listen to music, play games, watch videos? All of these things? It's really hard to imagine life without this technology. In the future, we'll be spending a lot more time in the cloud!

MANY PEOPLE PAY MONEY FOR SEVERAL STREAMING SERVICES TO WATCH DIFFERENT KINDS OF CONTENT, INCLUDING MOVIES, TV, AND SPORTS.

A TIMELINE OF THE CLOUD

1962 Joseph Carl Robnett Licklider writes about his vision of a global computer network.

1981 The IBM company launches, or starts, its first personal computer.

1991 The first web page goes live on the Internet.

1997 Professor Ramnath Chellappa is the first to use the term "cloud computing."

1999 Salesforce, one of the first cloud-computing software companies, launches.

2006 Google Docs and Amazon Web Services are launched.

2008 Microsoft Azure, the company's cloud-based platform, launches.

2011 IBM SmartCloud and Apple iCloud are launched.

2020 The cloud-computing market is valued at over $371 billion.

2021 Amazon Web Services controls 33 percent of the cloud-computing market.

A ZETTA-WHAT?

A company called Cybersecurity Ventures, which studies the Internet, estimates, or guesses, that the amount of data stored in the cloud will amount to about 100 zettabytes by 2025. A zettabyte is one sextillion bytes, or 1,000,000,000,000,000,000,000 bytes.

GLOSSARY

access: To have the right or ability to use something.

app: Short for application. Computer and smartphone programs that are made to perform certain functions.

billion: 1,000 million, or 1,000,000,000.

byte: A unit of computer data, made up of eight bits, often standing for a letter or number.

electromagnet: A piece of metal that becomes magnetic when an electric current is passed through or near it.

flash drive: A small storage device that can be plugged into a port on a computer, used to move files from one computer to another.

modem: A device that transforms data into signals that can be sent through telephone lines between computers, and is often involved in internet connections.

pandemic: An illness that is widely spread over a nation or the world.

predict: To guess what will happen in the future based on facts or knowledge.

resource: A usable supply of something.

router: A machine designed to send data from one place to another within a computer network or between computer networks.

satellite: A machine sent into space that moves around Earth, the moon, the sun, or a planet to complete scientific studies or a type of communication.

technology: A machine, piece of equipment, or method created by science, engineering, and other industries to invent useful tools or to solve problems.

FOR MORE INFORMATION

BOOKS

González, Echo Elise. *The Internet.* Chicago, IL: World Book, 2020.

Idzikowski, Lisa. *Connecting a Computer System.* New York, NY: PowerKids Press, 2019.

Rarus, Pat. *Careers in Info Tech.* San Diego, CA: ReferencePoint Press, 2020.

WEBSITES

Online Safety
kidshealth.org/en/kids/online-id.html#cathouse
Read reminders about how to stay safe on the internet.

A Short History of the Internet
www.scienceandmediamuseum.org.uk/objects-and-stories/short-history-internet
If you're curious about how the internet progressed from ARPANET, check out this site.

What Is the Cloud?
edu.gcfglobal.org/en/computerbasics/understanding-the-cloud/1/
Read more about cloud computing on this user-friendly site.

INDEX

APPS, 14, 15, 19, 20, 21, 23
ARPANET, 13
ARTIFICIAL INTELLIGENCE (AI), 26, 27
BYTES, 5, 29
COVID-19 PANDEMIC, 22, 23
DATA CORRUPTION, 10, 11
DATA ENCRYPTION, 24, 25
EXTERNAL DRIVES, 8
GAMING, 7, 10, 11, 23, 28
GOOGLE CLASSROOM, 23
HIGH-SPEED INTERNET, 14, 15
HYBRID CLOUD, 16
INTERNET SERVICE PROVIDERS (ISPs), 15
LICKLIDER, JOSEPH CARL ROBNETT, 12, 13, 29
LOCAL STORAGE, 6, 7, 10, 15
MCCARTHY, JOHN, 12
NETWORKS, 12, 13, 15, 16, 17, 19, 29
PERSONAL COMPUTERS (PCs), 8, 9, 10, 16, 29
PRIVATE CLOUD, 16
PUBLIC CLOUD, 16
SERVERS, 6, 7, 8, 14, 16, 17, 18, 19
SMARTPHONES, 4, 6, 14, 15, 19
SOFTWARE, 15, 18, 19, 29
STREAMING, 7, 11, 28
SYNCHRONIZATION, 19
TABLETS, 4, 6, 14
ZOOM, 23